AI AWARENESS SERIES

AI in Customer Service

Darian Batra

Contents

Introduction

Customer service is undergoing a profound shift. In an era where speed, personalization, and 24/7 availability are no longer "nice to have" but expected, artificial intelligence is becoming a core enabler of modern support operations. From chatbots and voice assistants to smart routing, quality monitoring, and predictive staffing, AI is redefining what customer service can look like—both for the customer and the teams behind the scenes.

This book, AI in Customer Service, is part of the AI Awareness Series—a collection designed to make artificial intelligence accessible, actionable, and responsible across key industries and business functions. Here, we focus on how AI is transforming the customer service landscape, not with abstract theory, but with practical tools, use cases, and strategic insight.

Whether you're a support leader, operations manager, CX strategist, or technologist, this book is written to help you understand how AI can be thoughtfully applied to improve service quality, enhance customer satisfaction, and increase team efficiency. Each chapter addresses a specific area of transformation—from personalization and automation to compliance, coaching, and ethical considerations.

- You'll learn how AI is being used to:
- Streamline ticket handling and automate routine tasks
- Deliver real-time personalization and proactive service
- Monitor quality and support agent performance at scale

Introduction

- Forecast demand and plan workforce capacity more accurately
- Analyze customer sentiment and feedback through Voice of the Customer tools
- Build AI-ready support teams and align technology with data strategy
- Balance innovation with fairness, privacy, and transparency

Above all, this book encourages a human-centered approach to customer service innovation. AI is not here to replace the human touch—it's here to enhance it. When implemented thoughtfully, AI can reduce friction, free up agents for more meaningful interactions, and help organizations create customer experiences that are not only faster and smarter, but also more empathetic and effective.

Let's begin.

Chapter 1: The Role of AI in Modern Customer Service

Let's begin by defining AI in the context of service transformation. Artificial intelligence refers to machines mimicking human abilities like learning, reasoning, and problem-solving. In service transformation, AI is used to automate and improve both customer interactions and operational processes, making services more efficient and responsive.

There are several key drivers pushing service organizations to adopt AI. First, AI improves efficiency by streamlining operations and automating tasks. Second, it allows for personalized customer experiences, which are critical for meeting growing expectations. Third, it helps reduce operational costs by optimizing how resources are used. And finally, the increasing availability of data makes AI a powerful tool for gaining insights and staying competitive.

AI has enabled new types of service models. Chatbots and virtual assistants offer real-time customer support, making interactions smoother and faster. Predictive analytics uses AI to forecast trends and personalize services based on data insights. And automated workflows help reduce manual work, making service delivery quicker and more consistent.

Chapter 1: The Role of AI in Modern Customer Service

Customer service has evolved significantly, moving from manual processes to automated ones. Automation introduced faster and more convenient service options. IVR systems let customers handle issues over the phone without speaking to a human. And online self-service portals gave customers the ability to manage services on their own, improving accessibility.

There have been several key milestones in customer service technology. Call centers made it possible to handle customer service at scale. CRM systems centralized customer data for better personalization. Automated ticketing systems improved how customer requests were managed. And today, AI-powered chatbots allow for instant, intelligent customer interactions around the clock.

From past transitions, we've learned some important lessons. Balancing advanced technology with a human touch is critical. Customers still want empathy alongside efficiency. Designing user-friendly interfaces ensures that technology supports, rather than hinders, the customer experience. And continuously improving processes helps organizations keep up with changing customer needs.

When comparing AI-driven automation with traditional rule-based automation, the differences are clear. AI can adapt and learn from data, handling complex and unstructured tasks. In contrast, rule-based automation follows predefined instructions and works best for repetitive tasks. Both have their place, but AI opens up possibilities that rule-based systems can't handle.

Let's look at some specific use cases that highlight AI's unique advantages. Natural language processing allows chatbots to understand and respond in human language. Predictive analytics helps businesses anticipate customer needs. And AI can offer personalized recommendations, increasing both engagement and satisfaction.

Of course, both approaches come with challenges. Rule-based systems struggle with flexibility and can't easily adapt. AI systems need large amounts of data and expert knowledge to work effectively. Integration into existing systems can be complex. And, most importantly, maintaining customer trust and transparency is critical when using these technologies.

AI maturity models help organizations understand their progress with AI adoption. They typically follow stages: starting with initial experimentation, moving to pilot projects, scaling up successful pilots, and finally achieving full integration where AI becomes a core part of operations. Each stage builds on the previous one.

Before integrating AI, organizations need to assess their readiness. High-quality data is essential for good AI outcomes. Leadership must actively support AI initiatives. Skilled talent in AI and data science is critical. And organizations need strong change management to adapt workflows and company culture for AI integration.

AI has a significant impact on customer experience metrics. It can improve customer satisfaction by personalizing interactions. AI helps achieve higher first contact resolution by giving agents real-time insights. It reduces response times with automation and improves the customer effort score by making processes easier and more seamless.

To measure AI's impact on customer satisfaction, organizations use a mix of methods. Customer surveys gather direct feedback. Analyzing behavioral data reveals patterns in how customers interact with AI. And sentiment analysis helps understand customer opinions and loyalty based on what they say and how they say it.

AI has also introduced new customer experience metrics. AI accuracy tracks how well AI systems perform. The automation rate shows the percentage of interactions handled entirely by AI. And the chatbot containment rate measures how often chatbots resolve queries without needing human support. These metrics help organizations fine-tune their AI strategies.

Chapter 1: The Role of AI in Modern Customer Service

To wrap up, AI plays a transformative role in modern service organizations. By driving efficiency, enhancing personalization, and opening new service possibilities, AI is reshaping how services are delivered and experienced. Understanding its capabilities, challenges, and impact on customer experience is key for any organization looking to stay competitive in today's market.

Chapter 2: AI Technologies Powering Customer Support

AI adoption in customer service is transforming how companies operate. It automates routine tasks like answering simple inquiries, allowing human agents to focus on complex issues. AI provides round-the-clock support, meaning customers can get help at any time, regardless of location. It also personalizes customer interactions by analyzing data and tailoring responses, which enhances engagement and satisfaction. Because of these benefits, organizations are investing heavily in AI to improve service quality and reduce costs.

Implementing AI brings both advantages and challenges. On the positive side, AI improves efficiency through faster response times and cost reductions. However, integrating AI with existing systems can be technically demanding and requires expertise. There's also the need for thorough employee training so staff can effectively work with AI tools.

And importantly, organizations must prioritize data security and ensure their AI models operate fairly and without bias.

NLP, or Natural Language Processing, allows AI systems to process both text and speech inputs from customers. It parses grammar and syntax, helping the system understand the structure and meaning of what customers say. NLP also detects sentiment and identifies key information within communications, which helps create more accurate and helpful responses.

Natural Language Understanding, or NLU, takes things a step further. It doesn't just parse language — it interprets the context and the user's intent. This means AI can grasp what the customer really wants to achieve and tailor its responses accordingly. Contextual understanding ensures that customer service feels more natural and relevant.

NLP and NLU power various applications in customer service. Chatbots and virtual assistants use these technologies to understand and respond naturally to customer queries. Sentiment analysis is another important use case — it helps companies assess customer emotions and respond proactively to improve service quality.

Machine learning plays a crucial role in customer service through both supervised and unsupervised learning. Supervised learning uses labeled data to train AI models to categorize and predict customer requests accurately. On the other hand, unsupervised learning finds hidden patterns in data, helping identify customer segments and emerging trends without predefined categories.

Predictive analytics leverages historical data to forecast customer behaviors and preferences. By analyzing trends, AI can predict what customers might need or want, enabling personalized service and marketing strategies. It also allows companies to proactively identify and address potential issues before they escalate.

Machine learning also supports classification tasks, such as automatically routing inquiries to the right team, triaging urgent requests, and prioritizing workflows. These tasks help businesses respond more quickly and efficiently, improving the overall customer experience.

Large Language Models, or LLMs, are a major breakthrough in AI. They can generate coherent, natural-sounding responses, making interactions feel human-like. Because they understand context deeply, they ensure relevant and meaningful conversations with customers. This leads to better user engagement with AI-powered systems.

Generative AI has several practical applications in customer service. It can draft emails automatically, saving time and ensuring a professional tone. It helps generate FAQs, making information more accessible to customers. And it creates dynamic dialogue scripts, allowing for consistent and tailored communication during customer interactions.

When using generative AI, it's important to address ethical considerations. We must ensure content is accurate, fair, and unbiased. Content quality assurance processes need to be in place to review AI-generated outputs before they reach the customer, maintaining trust and service standards.

Computer vision and OCR, or Optical Character Recognition, automate the extraction of information from documents. OCR technology converts scanned images or text into editable, searchable data quickly and with high accuracy. This reduces manual data entry, boosts efficiency, and minimizes errors.

OCR has multiple use cases in customer service. It enables accurate customer identity verification, helping prevent fraud. It automates form processing, reducing manual workloads. And it streamlines invoice management, ensuring accuracy and timeliness in financial workflows.

AI technologies like computer vision and OCR often integrate with other AI systems for seamless workflows. For example, combining OCR with NLP and machine learning allows for end-to-end automation — from capturing documents to resolving customer issues. Integration with backend systems ensures everything works together efficiently, enhancing overall customer service operations.

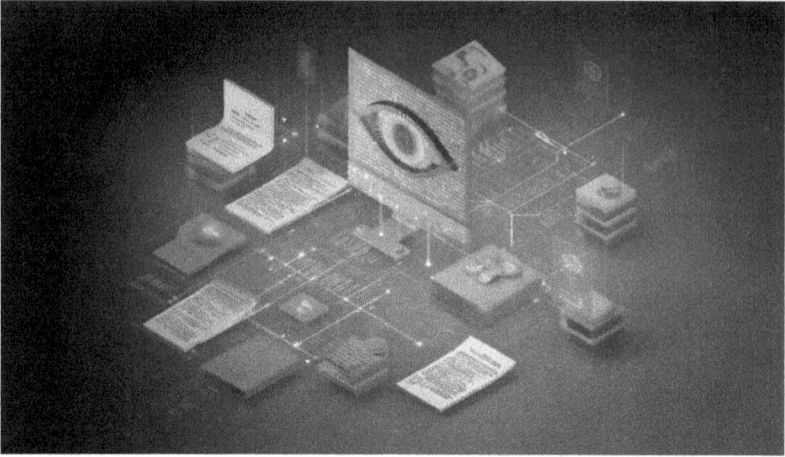

APIs — or Application Programming Interfaces — are essential for connecting different systems. They provide standardized interfaces that allow AI tools to integrate smoothly with existing customer service platforms. APIs enable seamless data and functionality flow, making operations more efficient and responsive.

Cloud platforms play a vital role in supporting scalable AI services. They offer flexible resources that can scale up or down based on demand. Cloud infrastructure also handles the large volumes of data needed for AI training and deployment. Plus, it supports real-time AI-powered customer interactions on a global scale.

Integration layers connect AI tools with backend systems, creating automated workflows that span the entire customer service process. They orchestrate how different AI components and backend services work together, improving the speed and accuracy of service delivery. This leads to a more seamless and efficient customer experience.

In conclusion, AI is reshaping customer service by automating routine tasks, enhancing personalization, and improving efficiency. From

natural language understanding to machine learning, generative AI, computer vision, and robust infrastructure — each technology plays a critical role. Understanding and leveraging these technologies will help organizations deliver better customer experiences and stay competitive in the evolving digital landscape.

Chapter 3: Personalisation and Proactive Service

When designing a chatbot, we start by deeply understanding user needs — this ensures the chatbot actually meets their expectations and provides value. Next, we select the right interaction channels, whether that's a website, app, or messaging platform, so the chatbot reaches users in the most convenient way. We also define clear use cases to focus the chatbot on specific tasks, keeping it purposeful and effective. Finally, integrating the chatbot seamlessly with existing systems means it can access the right data and support business processes without disruption.

Balancing automation with human interaction is crucial. While automation handles routine queries efficiently, human support must be available when conversations become complex. A well-designed chatbot knows its limits — it escalates to a human when necessary to ensure users always get the right level of support.

Users expect chatbots to respond quickly and accurately. They also expect the chatbot to understand the context of their conversation, so responses stay relevant. By managing user expectations — for example, by explaining what the chatbot can and can't do — we build trust and avoid frustration.

Chatbots should mimic natural conversation. This means structuring interactions logically, using simple and clear language, and having effective strategies for when things go wrong. If a user makes a mistake or the chatbot doesn't understand, it should guide the user back on track without confusion or dead-ends.

Chatbot design must be accessible and inclusive. This includes adding support for screen readers, voice commands, and other assistive technologies, as well as being mindful of cultural differences and offering multiple language options. This ensures the chatbot serves a diverse audience effectively.

A chatbot must accurately identify user intents — this is key to delivering useful responses. By categorizing different types of intents, we can organize the chatbot's responses effectively. We train natural language understanding models with varied examples to improve their ability to recognize intents, and we continually refine our intent definitions to make sure the chatbot adapts as user needs evolve.

Dialogue trees map out possible conversation paths. By visually mapping dialogue pathways, we ensure the chatbot can handle different user inputs logically. These trees need to be flexible to adapt to varied responses and scalable to handle more complex scenarios as the chatbot evolves.

Ambiguous queries are a challenge, but we can manage them by asking clarifying questions, offering users multiple response options, and using clues from previous parts of the conversation to interpret meaning. These techniques help the chatbot steer users toward the right answers even when the input is unclear.

To maintain a coherent conversation, chatbots must track user inputs and manage conversation history. Techniques like slot filling — where the chatbot collects key pieces of information — and entity recognition — where it identifies important elements in a user's message — help ensure relevant and accurate responses.

Memory is a powerful tool for personalization. Short-term memory allows a chatbot to keep track of information within a session, while long-term memory stores data across sessions to create a more personalized experience. Used well, memory can make interactions feel more tailored and engaging.

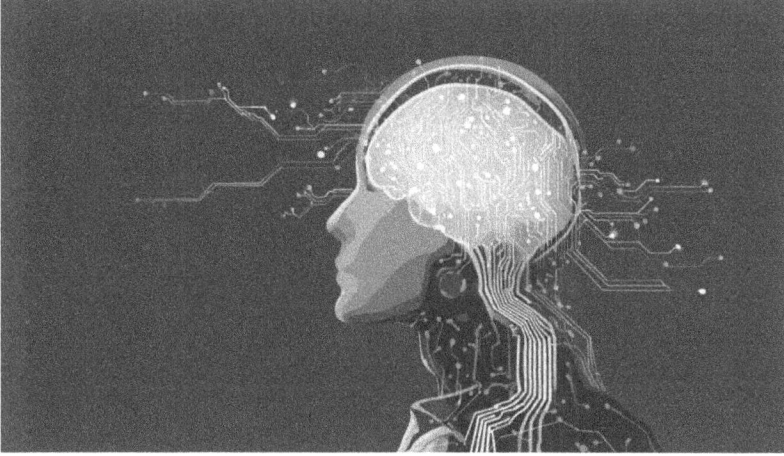

Handling user data responsibly is critical. Chatbots must comply with privacy regulations, anonymize sensitive data when possible, and always obtain user consent for data collection. Secure data storage protects information from breaches, building user trust and ensuring ethical data management.

To wrap up, we've covered the core elements of designing effective chatbots — from understanding user needs and managing conversations naturally, to modelling intents and safeguarding user data. By applying these principles, you'll be able to create chatbots that provide meaningful, engaging experiences while respecting user privacy and ensuring seamless interactions.

Chapter 4: AI in Quality Monitoring and Coaching

LLMs have a growing role in customer service. They're powering chatbots that give instant, consistent answers across platforms. Virtual assistants also use them to create more natural, helpful interactions. And when it comes to email, LLMs can automatically generate replies, reducing response times and freeing up human agents for more complex tasks.

Integrating LLMs brings some key benefits. They help scale customer service operations and ensure availability around the clock. But there are challenges too—keeping responses accurate, making sure the model understands context, and managing errors or misunderstandings. All of these need to be carefully addressed to create positive customer experiences.

Real-world case studies show the impact of LLMs. They boost customer satisfaction by delivering quicker, higher-quality responses. They also cut operational costs by automating routine tasks. And they streamline support workflows, making life easier for both customers and support teams.

When using LLMs for customer service, prompt design is crucial. Clear prompts help the model give accurate answers. Providing context makes the responses more relevant. And prompts need to handle diverse customer intents and language styles so interactions feel natural and inclusive.

Optimizing prompts is an ongoing process. By refining how we phrase prompts, we can make the model's answers clearer and more precise. This leads to fewer misunderstandings and more relevant responses—ultimately improving customer satisfaction and trust.

Prompt strategies need continuous testing. Customer needs change, so prompt designs must adapt too. Gathering feedback helps us refine prompts over time, while tracking performance metrics ensures we're delivering the best possible service in every interaction.

Ambiguous questions are common in customer service. Detecting and interpreting them correctly is key to avoiding misunderstandings and ensuring the customer feels heard and supported.

To handle vague inputs, we can ask follow-up questions or request more details. These techniques help clarify the customer's needs and allow the LLM—or human agents—to provide more accurate and helpful responses.

Sometimes, automation isn't enough. That's why it's important to have clear escalation protocols that hand over complex issues to human agents when needed. This helps maintain high service quality and builds customer trust by ensuring their concerns are addressed personally and effectively.

To protect customers, we need to set up content and safety filters. These filters prevent inappropriate or harmful outputs from reaching users and ensure the AI's behavior aligns with company policies and customer expectations.

Fairness is another critical aspect. We must be aware of potential biases in the data used to train LLMs and apply strategies to reduce those biases. This not only promotes ethical AI use but also helps maintain the brand's reputation and customer trust.

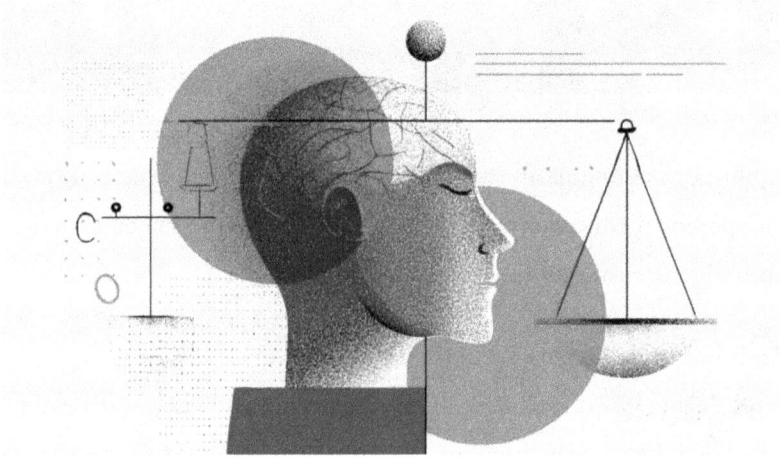

Transparency and accountability are essential. We should openly disclose when AI is being used in customer interactions. Having clear accountability measures ensures responsible deployment, while compliance with regulations protects both the business and customer rights.

To wrap up, leveraging LLMs in customer interactions holds huge potential—but it requires thoughtful implementation. By focusing on effective prompt design, handling ambiguity, enforcing ethical guidelines, and maintaining transparency, we can harness the power of LLMs to enhance customer service in a responsible and impactful way.

Chapter 5: Voice of the Customer (VoC) with AI

Multichannel interfaces allow users to interact with services through various platforms — like mobile apps, websites, or social media — giving them a seamless experience no matter the device. On the other hand, multimodal interfaces go a step further by allowing users to engage through different modes simultaneously — such as voice, touch, or visual inputs — making the interaction richer and more dynamic.

In today's digital world, customers expect seamless experiences. Whether they're switching devices or switching between input methods, they want interactions to feel natural and intuitive. Multichannel and multimodal interfaces aren't just convenient — they're becoming essential for engagement, accessibility, and user satisfaction.

Chapter 5: Voice of the Customer (VoC) with AI

Let's look at some real-world examples: Smart home systems blend voice commands with mobile apps and visual displays, allowing users to control devices effortlessly. Customer support centers now integrate voice calls, chat, and email — letting customers communicate on their terms. And social media platforms use text, images, video, and even voice to create engaging, multi-faceted user experiences.

AI has transformed how voice assistants and smart speakers operate. AI helps these devices understand natural language, even when phrased in different ways, making interactions feel more human. They also offer hands-free control — a big advantage for accessibility and convenience. And when linked to smart home systems, they act as a central hub for managing connected devices.

In chat platforms, conversational AI allows for more natural, flowing dialogue between users and systems. These AI-driven systems can answer questions, resolve issues, and even anticipate user needs, improving customer support and enhancing user engagement.

Chapter 5: Voice of the Customer (VoC) with AI

On social media, AI plays a key role in personalizing the user experience. By analyzing behavior patterns, AI helps platforms understand what users like and how they interact. This means users get more relevant content, whether that's posts, suggestions, or ads — making engagement higher and ads more targeted.

Visual interfaces, like image recognition, allow devices to understand and process visual data. For example, your phone recognizing faces in photos, or apps that scan products and give information instantly. Augmented reality takes it further by layering digital content over the real world — creating engaging, immersive experiences for users.

Voice input provides real benefits for customer experience. It allows users to interact without needing to use their hands — perfect for multitasking or when hands-free is safer or more convenient. Voice also opens up accessibility for those with mobility challenges, making digital services more inclusive. And by simplifying interaction, it increases user engagement by making tasks feel quicker and easier.

When we blend visual and voice inputs, we create interactions that feel more natural and intuitive. For example, a user might ask a question by voice and then get visual results on a screen. This combination also adapts to different contexts — whether someone is driving, walking, or sitting at a desk — giving them a flexible, user-centered experience.

Accessible interface design starts with key principles: Making sure information is perceivable — so it can be understood in different ways, Ensuring interfaces are operable by people with various abilities, And designing content to be understandable and robust, so it works with a range of assistive technologies. These principles help create experiences that are usable by everyone.

AI can open up new possibilities for users with disabilities. For example, AI-powered screen readers, voice commands for control, or visual recognition tools that assist people with visual impairments. These applications can empower users and create more equitable access to technology.

Of course, AI doesn't come without challenges. Bias in AI models, privacy concerns, and design limitations can all create barriers to fairness and trust. That's why following ethical guidelines is crucial — ensuring AI systems respect user rights and promote transparency. Inclusive design practices — like involving diverse user groups in testing — help make AI tools usable and fair for all communities.

To sum up: Multichannel and multimodal interfaces, especially when combined with AI, are reshaping the way users engage with digital platforms. They offer richer, more seamless experiences while enhancing accessibility and inclusion. But it's equally important to be aware of the ethical and design challenges — and to work towards solutions that benefit all users.

Chapter 6: AI in Ticket Management

Let's begin with an overview of case classification in customer service. Case classification is about organizing customer inquiries to make resolution faster and more efficient. By classifying cases based on their nature, urgency, and complexity, businesses can ensure that each case is prioritized correctly and sent to the right team. The main benefit here is improved response times and higher customer satisfaction, since cases are handled by the most suitable people from the start.

Traditional case routing has several challenges. Manual processes slow things down and make handling cases inefficient. Rule-based systems can only follow predefined logic, which often lacks the flexibility needed for real-life situations—leading to mistakes in routing and classification. All of this negatively affects customer satisfaction, as customers experience delays and possibly even misdirected responses.

This is where automation steps in. Automated systems improve both efficiency and accuracy by removing manual steps and reducing human error. They make the process faster and more reliable, ultimately enhancing the customer experience.

Now, let's look at predictive intent detection. By analyzing customer messages, systems can understand what the customer actually wants. Once intent is classified, cases can be automatically routed based on that predicted intent. This leads to quicker, more relevant responses— and a better service experience for the customer.

Techniques like natural language processing and machine learning are the backbone of intent analysis. These methods allow systems to interpret customer language more effectively and classify intents accurately.

Let's look at a few examples of how predictive intent detection is used. In chatbot systems, it helps the bot understand user queries and respond more personally and efficiently. Automated ticketing systems use intent detection to sort and prioritize tickets, speeding up resolution. And in customer support platforms, predictive intent enables tailored interactions, which boosts overall customer satisfaction.

Next, let's talk about skill-based routing. This approach directs cases to agents with the right expertise, ensuring that customer issues are handled by people who are best equipped to resolve them. The result is higher quality resolutions and greater customer satisfaction.

Sentiment analysis can also be a powerful tool in routing decisions. By understanding the customer's emotional state, systems can route sensitive cases to agents who are skilled in handling high-emotion situations with empathy.

Combining skill-based and sentiment-aware routing provides optimal case handling. Skill-based routing ensures technical expertise. Sentiment-aware routing matches emotional needs with the right agents. Together, these methods improve resolution outcomes and create more satisfying experiences for customers.

Now, let's explore priority scoring with machine learning. Priority scoring models look at multiple data points to assess how urgent or impactful a case is likely to be. This allows teams to make better decisions about which cases to address first—improving both response times and service quality.

Several key features help machine learning models with prioritization. Customer history reveals patterns that inform urgency. Classifying case types helps categorize cases correctly. Sentiment analysis highlights emotional cues linked to urgency. And considering the interaction channel—whether it's email, chat, or phone—adds more context for better decisions.

Automated priority scoring improves service outcomes in three ways: It reduces response times for high-priority cases. It boosts customer satisfaction by ensuring timely responses. And it supports intelligent workload management by helping distribute work efficiently among support teams.

To wrap up, automated case classification and routing bring real benefits to customer service operations. By using predictive intent detection, skill and sentiment-aware routing, and machine learning for priority scoring, businesses can respond faster, improve customer satisfaction, and manage workloads more intelligently. Ultimately, these advancements

Chapter 7: Demand Forecasting and Workforce Planning

Agent Assist and Co-Pilot technologies are AI-powered tools designed to support agents with timely and relevant information. They provide helpful suggestions and automate routine tasks, making agents more efficient and accurate. By assisting during customer interactions and business processes, they help deliver better overall service and enhance customer relationships.

These technologies bring clear benefits to customer support, sales, and business operations. They boost productivity by reducing manual tasks, allowing staff to focus on higher-value work. AI-driven insights improve decision-making, and faster response times enhance customer satisfaction while streamlining operations.

Agent Assist and Co-Pilot integrate smoothly with existing workflows, minimizing disruption. This seamless integration ensures business processes continue uninterrupted while still enhancing agent productivity. By embedding into current systems, they optimize workflows and contribute to better outcomes for both agents and customers.

Real-time knowledge engines work by continuously analyzing live interactions to understand context and needs. They deliver timely, relevant information that helps agents make better decisions in the moment. By tailoring recommendations to each specific interaction, these tools maximize their usefulness and effectiveness.

In customer interactions, these AI engines provide real-time suggestions that improve agent efficiency and accuracy. They offer troubleshooting guidance, reducing resolution times and enhancing outcomes. Ultimately, this leads to higher customer satisfaction and boosts agents' confidence in handling complex queries.

Live translation supports real-time communication between people speaking different languages. It helps break down language barriers, enabling smoother global interactions. This AI-powered translation improves the quality of customer-agent interactions by making multilingual support seamless and efficient.

AI can also summarise conversations in real time, helping agents quickly review key points and next steps. These summarised insights streamline follow-ups and ensure that nothing important gets overlooked in customer interactions.

By enabling service to diverse audiences, live translation and summarisation enhance global customer engagement. They help reduce friction in customer interactions, making engagement smoother and more effective. This results in increased customer satisfaction and loyalty from a broader, worldwide customer base.

Smart macros are a key tool for automating routine tasks like data entry and standard replies. By handling these repetitive actions automatically, smart macros save time and allow agents to focus on more complex and valuable tasks.

AI-assisted drafting helps agents by generating response drafts based on the context of each conversation. This improves both the speed and consistency of responses. While AI drafts the replies, agents still review and personalize them before sending, ensuring high-quality communication.

Smart macros and AI drafting tools together improve both productivity and consistency for agents. By automating routine tasks and supporting high-quality response drafting, they enable clear, consistent communication across the team. This combination ensures better efficiency and professionalism in all customer interactions.

AI plays a vital role in monitoring communication for regulatory compliance. It scans interactions in real time, detecting potential compliance risks. This continuous monitoring helps organizations ensure they're following legal and policy requirements.

When compliance risks are detected, AI systems send real-time alerts to the right people for quick action. They also offer actionable recommendations to help agents and supervisors take corrective measures effectively, reducing the risk of violations.

AI helps minimize compliance risks and ensures that organizations adhere to policies. By monitoring and managing communications, AI supports consistent, controlled, and compliant interactions across all channels. This proactive approach helps prevent errors and regulatory breaches before they happen.

In summary, Agent Assist and Co-Pilot experiences are powerful tools for enhancing productivity, improving decision-making, and ensuring compliance. By integrating seamlessly into workflows, providing real-time support, and automating key tasks, these AI-driven solutions empower agents and businesses to deliver better service and achieve more consistent outcomes.

Chapter 8: Building an AI-Ready Support Organisation

Sentiment and Emotion AI relies on powerful machine learning and natural language processing technologies. These systems analyze a variety of data—text, voice, even facial expressions—to identify people's feelings and attitudes. Multimodal emotion detection takes this a step further by combining cues from different sources to give a much more accurate reading of emotions, helping systems respond in more human-like ways.

Let's break down three key concepts. First, sentiment—this refers to the positive, negative, or neutral tone in communication. Next, emotion, which is more complex—things like joy, anger, or sadness. Finally, affect, a broader term that covers both sentiment and emotion, giving us a full picture of a person's emotional state in any given interaction.

To capture sentiment in live conversations, systems use natural language processing to analyze chat and voice transcripts in real time. Sentiment scoring algorithms then assign numerical values to detected sentiments, giving a measurable output. For even greater accuracy, multimodal sentiment analysis blends audio and visual signals with text to gauge emotions more precisely during interactions.

When integrated into customer service platforms, sentiment analysis provides real-time emotional insights. This lets agents adapt their responses as they interact with customers. It also helps agents become proactive—anticipating customer needs or reactions based on detected emotions—leading to better outcomes for both the customer and the service provider.

Real-time sentiment and emotion analysis offers some clear benefits—like improved customer satisfaction and quicker problem resolution. However, it comes with challenges too. The need for high-speed processing to analyze live data streams is a major hurdle. Plus, ensuring data privacy and maintaining analysis accuracy—especially across different languages and contexts—remains a critical concern.

Emotion detection works across text, voice, and video. In text, systems analyze word choices to detect emotions. Voice brings out emotional cues like tone and pitch. Video goes even further by capturing facial expressions and body language. By combining these—what we call multimodal emotion detection—we get a deeper, more reliable understanding of a person's emotional state.

To make the most of emotion detection, data needs to be integrated across multiple channels. Multi-channel data fusion brings together insights from text, voice, and video, creating a consistent, holistic view of user emotions. This approach overcomes the limitations of relying on a single channel, allowing for much richer emotional insights.

Let's look at some real-world uses. In healthcare, emotion classification supports mental health monitoring and personalized care. Retail businesses use it to tailor experiences and improve customer satisfaction. And in entertainment, understanding audience emotions helps deliver content that resonates and engages better. These industries show how valuable emotion AI can be in practice.

Escalation protocols are rules that trigger higher levels of support when negative sentiments or emotions are detected. By monitoring affective signals—like frustration or anger—systems can determine when it's time to escalate an issue. This ensures that serious concerns are addressed quickly, improving the overall quality of service.

For these escalation triggers to work, customer support workflows need real-time monitoring tools. These tools detect emotional signals during interactions. Once detected, agents are alerted through internal systems, helping them react quickly. In some cases, workflows are even automated to route customers to supervisors or specialists when needed.

Affect-aware escalation has a real impact on customer service outcomes. It helps resolve issues faster, reducing downtime for the customer. This responsiveness boosts user satisfaction and increases the efficiency of the support team. In the long run, it also helps with customer retention by ensuring problems are handled with care and speed.

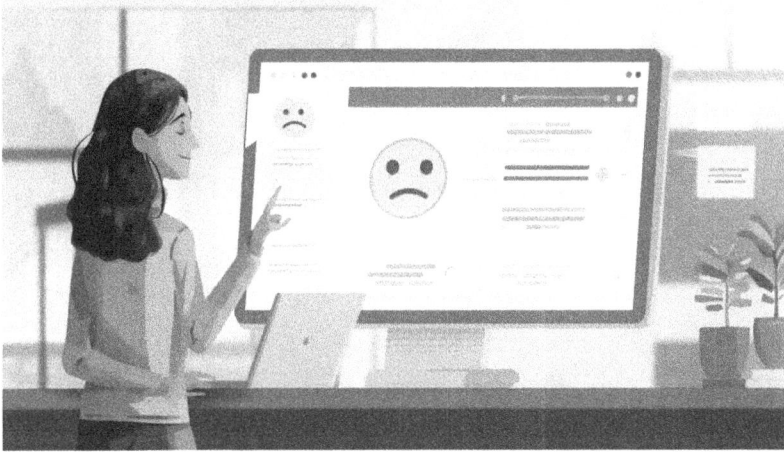

To wrap up, Sentiment and Emotion AI is transforming how businesses engage with customers in real time. By tracking emotions, analyzing data across channels, and setting smart escalation triggers, organizations can provide faster, more personalized support. While challenges like processing speed and data privacy remain, the benefits to customer satisfaction and service quality are undeniable.

Chapter 9: Data Strategy for AI in Support

Voice of Customer, or VoC, is all about systematically gathering customer expectations, preferences, and feedback. By listening closely to our customers, we can better align products and services to meet their needs. When done right, VoC not only improves satisfaction but also builds long-term loyalty—making it a critical part of any customer experience strategy.

Experience mining takes customer experience management a step further. It analyzes user behaviors and interactions to uncover patterns and root causes behind customer sentiments. When we integrate VoC with experience mining, we unlock actionable insights that enable smarter, data-driven decisions aimed at improving the overall customer experience.

To collect meaningful text and voice feedback, companies often use structured surveys for direct input. Social media platforms are another rich source of real-time, unfiltered customer opinions. And of course, call centers and chatbots provide channels for both voice and text interactions—giving us even more data to analyze and act upon.

Natural Language Processing, or NLP, helps us make sense of all this feedback. Tokenization breaks down text into smaller units for easier processing. Entity recognition extracts important details like names or locations from text. And sentiment analysis gauges the emotions behind feedback, giving us valuable insights into how customers feel.

With these techniques, we can extract key elements from customer feedback. Sentiment analysis tells us how customers are feeling—positive or negative. Intent recognition reveals what customers are hoping to achieve or communicate. And theme extraction helps us spot common issues or recurring topics—allowing for targeted actions based on real customer priorities.

Unsupervised learning is a powerful way to discover topics within customer feedback without predefined categories. One common method is Latent Dirichlet Allocation, or LDA, which automatically finds hidden topics in large text datasets. This saves time and effort by letting algorithms find patterns for us—making topic discovery scalable and efficient.

Clustering algorithms help segment feedback into meaningful groups. K-means clustering splits data into distinct clusters based on similarities. Hierarchical clustering creates nested groupings without needing to set a cluster number in advance. Both approaches help us segment feedback for more focused analysis and sharper insights.

Monitoring trends over time lets us see how customer concerns evolve—and spot emerging issues before they escalate. By prioritizing actionable topics, we can respond proactively to customer needs and focus resources where they'll make the most impact.

Predictive analytics lets us spot patterns in historical feedback that could signal customer churn risks. By analyzing these patterns, we can predict which customers are likely to leave and take action early to retain them. This targeted approach boosts customer satisfaction and helps reduce overall churn rates.

Machine learning models can analyze a wide range of feedback to score customer satisfaction. They deliver nuanced insights—capturing subtle shifts in sentiment that might be missed otherwise. This lets us monitor experience quality on a granular level and act on insights before problems grow.

Predictive analytics has powerful applications for customer retention. It enables personalized engagement by tailoring communication to each customer's preferences. It also optimizes resource allocation—making sure support and marketing efforts focus on the right customers. By reducing churn and driving loyalty, predictive analytics becomes a key tool in any retention strategy.

AI can automate workflows for handling customer feedback—making sure the right teams get the right information fast. This automation increases efficiency and helps resolve customer issues more quickly—ultimately improving satisfaction.

Closed-loop systems allow for real-time adaptation based on immediate feedback. They enable dynamic personalization, adjusting offers or responses to fit each customer's situation. This level of responsiveness leads to more engaged—and ultimately more satisfied—customers.

To ensure continuous improvement, it's critical to monitor feedback outcomes consistently. Measuring the impact of interventions helps organizations refine their processes over time. With this approach, customer experience isn't just managed—it's actively enhanced on an ongoing basis.

To conclude, integrating Voice of Customer, experience mining, and AI-driven analytics creates a powerful toolkit for enhancing customer insights and driving smarter business decisions. By applying these advanced approaches, organizations can improve customer satisfaction, reduce churn, and foster long-term loyalty—turning insights into real impact.

Chapter 10: Quality Monitoring And Compliance At Scale

As support operations scale, maintaining consistent quality becomes critical. Scalable quality assurance helps ensure every customer receives a uniform experience, even as demand grows. It also supports regulatory compliance by maintaining standards across expanding teams. Ultimately, scalable QA improves efficiency and protects your brand's reputation during periods of growth.

Managing large support teams comes with challenges, particularly around maintaining consistent service quality. Differences in agent performance, communication styles, and knowledge levels can all impact the customer experience, making it hard to uphold consistent standards without scalable solutions.

Automation and AI bring powerful benefits to quality monitoring. With continuous monitoring, AI removes the risk of human bias and fatigue by objectively assessing interactions 24/7. This reduces repetitive manual work for staff, allowing them to focus on more strategic tasks. Plus, AI delivers valuable insights that not only enhance support quality but also ensure compliance with regulations.

AI evaluates conversations by analyzing content, tone, and adherence to protocols. Natural language processing enables AI to accurately assess the meaning and quality of conversations. It can also detect tone and sentiment, helping gauge the emotional context. Importantly, AI checks whether agents follow established protocols — and it can do this both in real time and after the interaction ends, ensuring continuous improvement.

When AI scores conversations, it uses key metrics like compliance adherence to ensure all responses meet legal and regulatory standards. It measures customer sentiment to assess satisfaction and checks response accuracy to ensure questions are answered correctly. Resolution effectiveness evaluates how well the issue was solved, and AI also assesses whether empathy was shown, capturing the human element of support.

Compared to manual reviews, AI scoring offers faster, more consistent evaluations. It removes human bias, delivering objective assessments across all reviews. Plus, AI supports continuous monitoring and detailed analytics, which help organizations target coaching efforts and drive ongoing improvement in support interactions.

Anomaly detection with AI means identifying unusual behaviors or patterns in support interactions. AI can pick up irregular language that deviates from normal patterns, monitor unexpected sentiment shifts, and flag breaches of protocol. These detections help surface potential issues before they escalate.

The anomalies detected by AI have significant implications. Compliance violations can result in fines or legal action. Identifying quality drops helps maintain high standards and customer satisfaction. Early detection of security risks protects against breaches, while spotting fraudulent activity prevents financial losses and protects the organization's integrity.

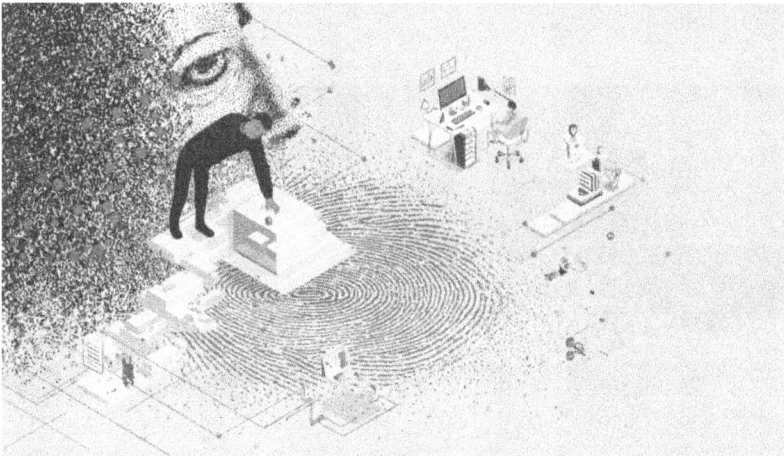

When AI detects anomalies, organizations can respond quickly. Immediate alerts notify supervisors for prompt action. Investigations can then be launched to find the root cause. Retraining agents ensures they're better prepared for future incidents, and AI models can be adjusted to improve future detection and reduce risks.

AI automates the detection of compliance breaches, making it easier to flag and address issues quickly. This automation supports compliance efforts across large-scale support operations.

Flagged compliance issues are automatically routed to the right teams for review. The escalation process ensures that potential violations are handled quickly and effectively, preventing small issues from turning into larger problems.

AI also helps protect sensitive data by automatically detecting personal identifiers and payment details. These are then redacted — or removed and masked — to prevent unauthorized access. This process is critical for maintaining privacy and ensuring compliance with data protection regulations.

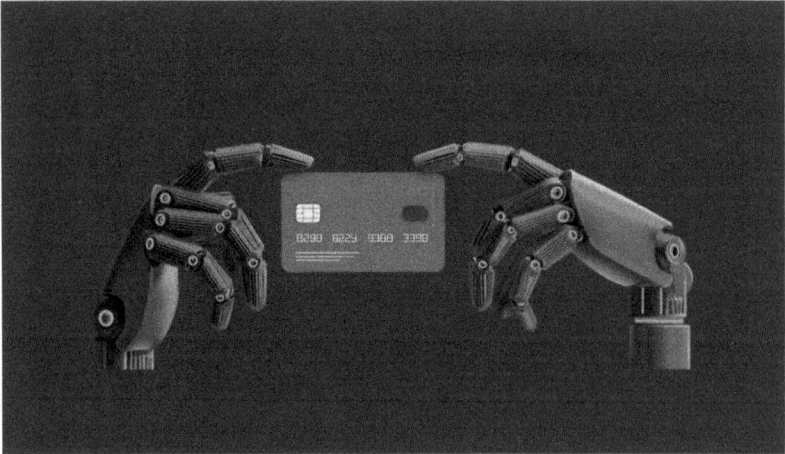

In summary, leveraging AI for quality monitoring and compliance allows organizations to scale their support operations confidently. It ensures consistent service, helps meet regulatory requirements, and protects both the customer experience and the organization's reputation. By integrating AI into quality assurance processes, companies position themselves for sustained growth and operational excellence.

Chapter 11: Workforce Analytics And Coaching Support

Workforce analytics has become essential in modern organizations. By analyzing data on employee productivity and effectiveness, companies can identify ways to boost performance. Attendance monitoring allows leaders to spot patterns and address absenteeism proactively. Engagement insights help organizations foster job satisfaction and retain top talent. Altogether, these analytics empower data-driven decision-making that benefits both employees and the organization.

AI plays a transformative role in workforce management. Automated data analysis means AI can handle large, complex datasets quickly, providing actionable insights without manual effort. Real-time monitoring enables managers to track workforce activities as they happen, allowing for immediate adjustments. Predictive analytics not only forecasts future trends but also supports personalized coaching to help agents perform at their best.

Agent performance dashboards powered by AI offer real-time data visualization, giving managers instant insights into how agents are performing. Customizable KPIs allow these dashboards to focus on the most relevant goals. Trend analysis and benchmarking help track progress over time, while built-in alert systems notify managers of potential issues before they escalate. These features make dashboards a powerful tool for performance management.

AI-driven insights take performance tracking to the next level. By detecting data patterns, AI identifies key trends and outliers in agent performance. Predictive analytics allows managers to anticipate challenges and plan ahead. And with proactive coaching support, managers can use these insights to tailor development strategies for each agent, leading to improved outcomes and higher engagement.

Both managers and agents benefit from these AI-enhanced systems. Managers gain a clearer picture of team performance, enabling them to provide better support and targeted coaching. Agents receive timely, constructive feedback that helps them grow and stay motivated. This combination drives continuous improvement and better overall results for the organization.

AI algorithms analyze vast amounts of data, including agent behavior and customer interactions, to deliver personalized coaching recommendations. By evaluating performance metrics, coaching becomes more effective and better suited to each agent's needs. This personalized approach maximizes the impact of coaching sessions and leads to meaningful improvements.

To be effective, feedback must be timely, specific, and constructive. Providing clear, actionable feedback soon after events ensures it remains relevant and impactful. With AI-driven insights, managers can offer more precise guidance during coaching sessions, ultimately leading to better development outcomes for agents.

Personalized coaching doesn't just enhance skills — it also boosts motivation. When agents feel supported with targeted feedback and development opportunities, they become more engaged in their work. This, in turn, leads to higher performance levels, better customer satisfaction, and improved retention for the organization.

Workload fluctuations can be caused by a variety of factors, including seasonal trends, promotional campaigns, and unexpected events. If not managed effectively, these fluctuations can lead to burnout or underutilization of agents, both of which negatively impact performance and morale. Understanding these challenges is the first step toward better workload management.

Predictive analytics uses historical data to identify patterns and anticipate future workload demands. By also considering external factors, such as market trends or upcoming events, AI models can deliver highly accurate forecasts. This allows organizations to proactively manage resources, ensuring they're prepared for workload changes before they happen.

AI-powered scheduling tools help managers ensure the right number of agents are available at the right times, optimizing staffing levels. By balancing workloads, AI prevents overwork and underutilization, supporting both productivity and employee well-being. Ultimately, smarter staffing leads to reduced costs, higher employee satisfaction, and improved service quality.

To conclude, leveraging AI in workforce analytics and coaching support offers a powerful way to enhance agent performance and management effectiveness. By using data-driven insights, real-time dashboards, personalized coaching, and predictive workload management, organizations can boost both operational efficiency and employee satisfaction.

Chapter 12: Automating Repetitive Workflows

Modern businesses face increasing pressure to operate efficiently. By streamlining routine tasks with automation, companies can save time and reduce the manual effort required across many processes. Automation also cuts down on human error, ensuring tasks are done accurately and reliably. And by automating workflows, businesses can boost productivity and deliver consistent outcomes, even in fast-changing markets.

Manual workflows often come with a range of challenges—these can include inefficiencies, errors, delays, and increased costs. Processes that rely too much on human intervention are prone to mistakes and bottlenecks. Over time, this can drag down performance and limit a company's ability to scale effectively.

Automating repetitive tasks brings several key benefits. First, enhanced accuracy—automation reduces human error and delivers consistent precision. It also improves employee productivity by freeing staff from mundane tasks, allowing them to focus on more strategic work. Cost reduction is another advantage, as automation saves time and resources. And finally, automation helps with compliance by ensuring that regulations and standards are followed consistently.

Robotic Process Automation, or RPA, uses software robots to mimic human interactions on digital systems. These bots can automate repetitive, rule-based tasks, improving efficiency and reducing manual work. A major advantage is that RPA can be deployed without changing your existing IT infrastructure, making it easy to adopt in most environments.

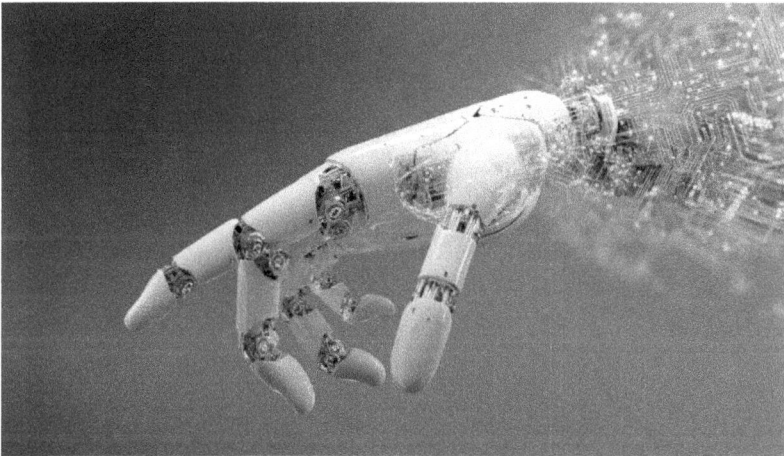

Artificial intelligence takes RPA to the next level. By enabling complex decision-making, AI allows automation to handle more advanced tasks. With natural language processing, RPA can understand and work with unstructured data like emails and documents. AI also gives RPA learning capabilities, meaning systems can improve over time. This makes it possible to automate dynamic workflows that involve changing data or unpredictable scenarios.

AI-triggered RPA is already making an impact across industries. In finance, it automates tasks like invoice processing and data management, improving accuracy and speed. Healthcare uses it for managing patient data securely, enhancing service delivery. And in customer service, AI-triggered RPA helps automate support tasks, reducing response times and improving the overall experience for users.

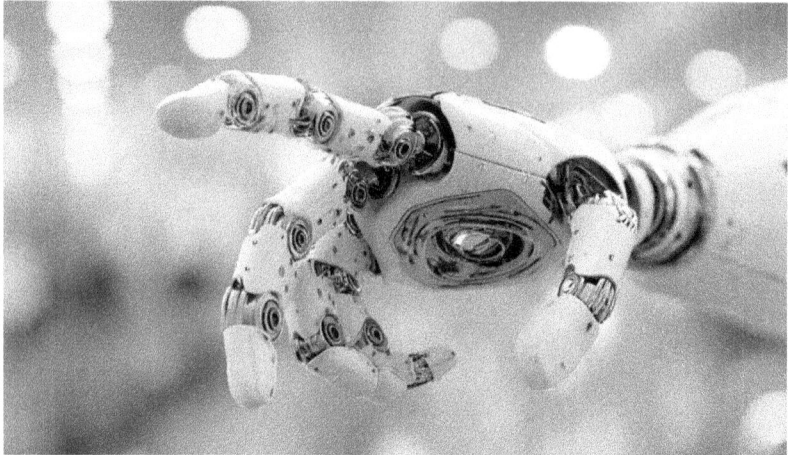

Backend task execution refers to automated processes that run in the background, triggered by user actions on the front end. These tasks operate without visible delays, ensuring a smooth user experience. They also enable real-time data processing, keeping systems updated instantly. Plus, backend tasks can integrate various systems, allowing them to communicate and work together seamlessly.

Here's how front-end actions trigger backend processes. When a user clicks a button or submits a form, backend workflows can start immediately. APIs are key—they connect front-end actions with backend automation for smooth data flow. And event-driven architectures make sure that backend systems respond right away to front-end events, improving efficiency.

To integrate backend automation effectively, some best practices include using well-designed APIs for clear communication between systems. It's also important to handle events precisely, so system interactions are quick and reliable. Strong error management is crucial too—it helps avoid failures and keeps systems stable. Examples like automated order processing or dynamic content updates show how this kind of integration can enhance both system performance and customer satisfaction.

To wrap up, automating repetitive workflows with AI-driven tools and smart integration strategies can transform business operations. By combining RPA, AI, and effective backend integration, organizations can boost efficiency, reduce errors, and free up human talent for higher-value tasks. Whether you're just starting or looking to optimize existing processes, these approaches offer powerful ways to streamline workflows and achieve better outcomes.

Chapter 13: Smart Knowledge And Faq Management

Smart knowledge systems are all about bringing together different types of information — both structured data like databases and unstructured data like documents — into one unified platform. AI helps to manage this knowledge by organizing, updating, and ensuring the information is both accurate and current. Finally, multi-channel delivery means your knowledge reaches users through various platforms, providing consistent and seamless access wherever they need it.

Effective FAQ management is essential because it reduces the overall support burden by providing instant answers to common questions. This doesn't just help your support team — it also boosts user satisfaction by giving people the right information at the right time. And, by providing reliable FAQ content, you help users solve problems quickly on their own, which can significantly improve their experience with your product or service.

AI-powered article generation transforms the way knowledge base content is created. By systematically extracting key information from various data sources, AI can draft articles much faster than manual methods. This automation speeds up the entire content production process while still maintaining high standards of quality and consistency.

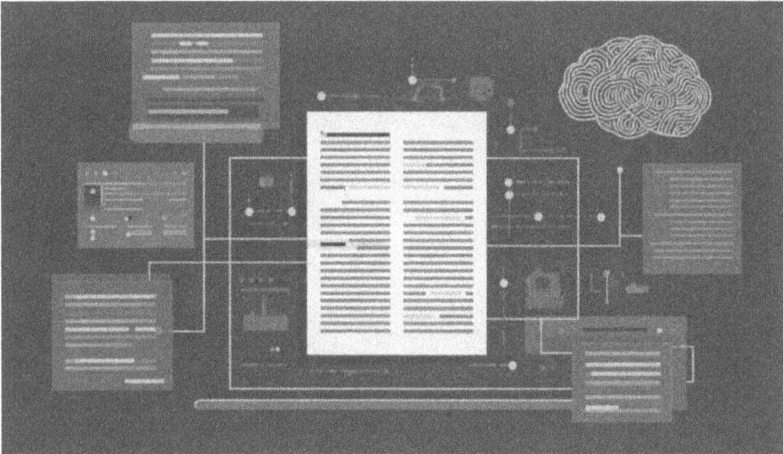

Natural Language Processing, or NLP, plays a key role in making sure the articles generated are clear and accurate. By understanding and refining language, NLP ensures that content is easy to understand and free of ambiguity — making the knowledge base far more useful to end-users.

AI doesn't just create content — it continuously improves it. By analyzing user feedback and interactions, AI can spot areas where content needs updating or where there are knowledge gaps. The content refinement process is ongoing, allowing the system to adapt to changes in information and user needs. With adaptive learning, AI systems evolve over time, ensuring that your knowledge base remains accurate and relevant.

AI enhances the user experience with intelligent suggestions. It does this by analyzing user queries and understanding the context behind them. Based on this analysis, the system can provide highly relevant suggestions in real time. This not only helps users find what they're looking for faster but also reduces the effort they need to put into searching — creating a smoother, more efficient experience.

AI algorithms can take things a step further by personalizing search results. By analyzing user behavior and search patterns, AI can tailor results to fit individual preferences. This ensures that users see content that's most relevant to them, improving both the quality of their search experience and their satisfaction with the knowledge base.

AI systems don't stop at making suggestions — they also monitor how well those suggestions are working. Through continuous performance monitoring, they can detect when search relevance drops or when certain results aren't performing as expected. By dynamically adapting relevance models based on this data, AI ensures that search results remain accurate and useful. This user-centered approach means the system evolves in line with user expectations.

One of AI's most powerful features is its ability to analyze user interactions and feedback to identify knowledge gaps. By studying queries, clicks, and interaction trends, AI can reveal common user difficulties. At the same time, analyzing feedback patterns helps to pinpoint where content is missing, outdated, or needs improvement — ensuring that your knowledge base keeps up with user demands.

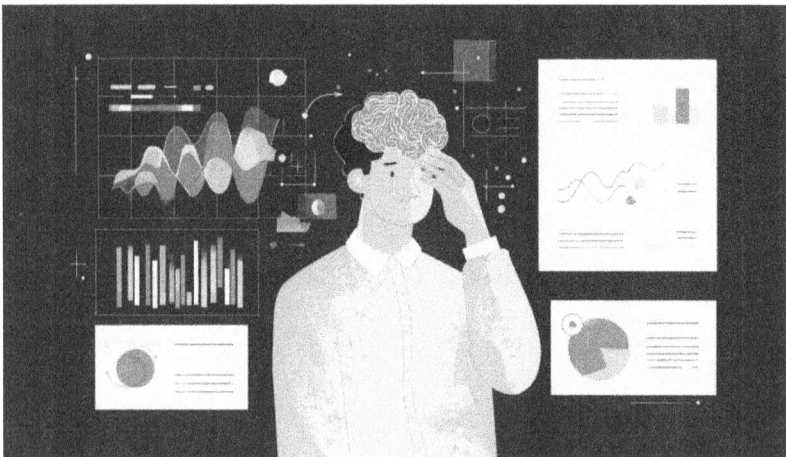

Detecting missing or outdated information is another critical function. AI can highlight areas where content is no longer relevant or where key information is absent. This proactive approach means you can address these issues before they impact the user experience.

AI-driven insights take the guesswork out of content strategy. By analyzing complex data, AI uncovers trends and patterns that inform smart content decisions. These insights help prioritize which articles need attention first, ensuring updates have the greatest impact. Aligning content updates with both user needs and business objectives keeps your knowledge base both useful and strategically focused.

To wrap up, we've explored how AI is transforming knowledge and FAQ management — from automated article creation and natural language processing, to personalized search and continuous improvement through feedback analysis. By leveraging these advanced AI tools, organizations can build smarter, more efficient knowledge systems that truly meet the needs of their users while also achieving business goals.

Chapter 14: Ethics And Bias Management

Ethical considerations in AI start with fairness. Ethical AI aims to minimize bias and promote equal outcomes for all users, regardless of their background. Transparency is also essential—users should understand how AI decisions are made, and developers must be accountable for those decisions. Lastly, ethical AI prioritizes user safety and ensures that AI behavior aligns with societal values and norms, protecting people from harm and misuse.

Customer-facing AI applications face unique challenges. They need to interact with users in real time, meaning responses must be instant and seamless to keep users engaged. These systems also serve diverse user demographics, so they must handle different cultural backgrounds, languages, and accessibility needs. Privacy is a big concern—customer data must be protected to maintain trust. And, of course, AI must minimize biases to ensure fair treatment and sustain customer confidence.

Let's look at common biases found in customer-facing AI. Data-driven biases often come from imbalanced samples or historical prejudices within training data. On the other hand, algorithmic design biases arise when the way an algorithm is built reflects hidden assumptions, leading to unfair AI behavior. These biases can significantly affect customer interactions, sometimes resulting in unfair or inaccurate outcomes.

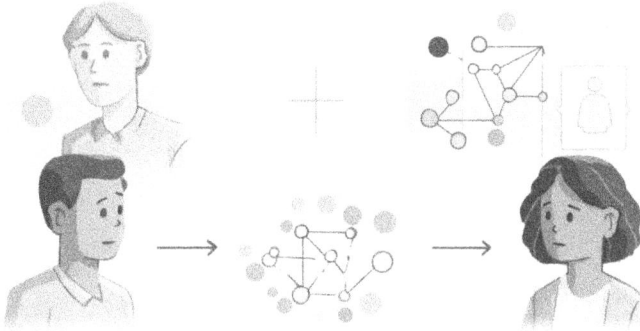

Bias in customer interactions can show up in various ways—like AI chatbots misunderstanding certain dialects or recommendation systems favoring one group over another. These examples highlight how bias can impact the quality and fairness of AI services, undermining the customer experience.

Bias directly affects user trust and satisfaction. When users perceive AI as biased, their trust in the system—and often in the company—declines. Negative perceptions can hurt your brand's reputation and reduce customer loyalty. Addressing bias isn't just about fairness; it's a key part of creating inclusive, reliable AI experiences that customers will trust and return to.

Now, let's discuss how we detect and mitigate bias. Fairness metrics help quantify whether AI models are treating different groups equitably. Data audits are crucial—they examine datasets for biased or unbalanced information. Simulation testing allows us to model different AI decision-making scenarios, helping spot potential bias before deployment.

On the algorithmic side, there are approaches specifically designed to reduce bias in AI models. These might include techniques like reweighting data, adjusting model parameters, or incorporating fairness constraints during model training.

Human-in-the-loop strategies are a powerful tool for ongoing bias management. Expert monitoring ensures that human reviewers can catch bias or errors in real time. Continuous review helps refine AI outputs, while adapting AI systems based on human insights ensures they remain relevant and fair in changing customer contexts.

Inclusive dataset curation starts with thoughtful data collection. It's important to represent diverse populations within your datasets. This means capturing a variety of contexts, environments, and situations so your data is truly reflective of the real world. A balanced dataset avoids both over-representation and under-representation, creating a fair foundation for AI models.

To identify and fix gaps in datasets, data profiling can help you spot missing or inconsistent segments. Assessing the impact of bias lets you recognize skewed data before it causes problems. And by integrating user feedback, you gain insights into real-world issues, which helps guide data improvements and corrections.

Documentation and transparency are key parts of responsible dataset management. Recording dataset provenance—the origin and history of the data—helps with accountability. Documenting your collection procedures highlights any biases that may have crept in during data gathering. Transparent documentation builds trust with stakeholders by clearly showing what the data can and can't do.

To wrap up, managing ethics and bias in customer-facing AI is critical for creating systems that are fair, transparent, and trusted by users. By understanding common biases, using effective detection and mitigation techniques, and curating inclusive datasets, we can ensure AI-driven customer interactions are ethical, reliable, and aligned with societal values.

Chapter 15: Privacy, Transparency And Regulation

Let's begin with an overview of GDPR principles. GDPR is all about protecting individuals' privacy by setting strict rules on how personal data is handled. The Data Minimization principle means only collecting data you really need. The Purpose Limitation principle ensures that data is only used for clearly stated, legitimate reasons. Together, these principles guide ethical and transparent AI use.

For AI systems to comply with GDPR, there are key requirements. First, lawful consent—AI must have permission before using personal data. Next, strong data security is critical. AI systems also have to support data subject rights like access and deletion. Finally, Data Protection Impact Assessments are essential when AI poses a high risk to privacy.

Handling customer data with AI comes with challenges. Bias in data can affect fairness and accuracy. Managing large volumes of data requires efficient technology. Transparency is crucial—not just for trust, but for ethical handling. Best practices like data anonymization and privacy-by-design help protect customer privacy while maintaining compliance.

Explainable AI is critical in customer-facing applications. It helps customers understand AI decisions, reducing confusion and uncertainty. This clarity not only builds trust but also ensures compliance with regulations. When customers understand how AI works, they're more likely to accept and trust its use in services.

There are several ways to improve model transparency. Using interpretable models makes decisions easier to understand. Feature importance analysis shows which data points most influence decisions. Local explanation methods like LIME and SHAP help clarify individual predictions. And visualization tools make it easier to explain complex AI processes to customers.

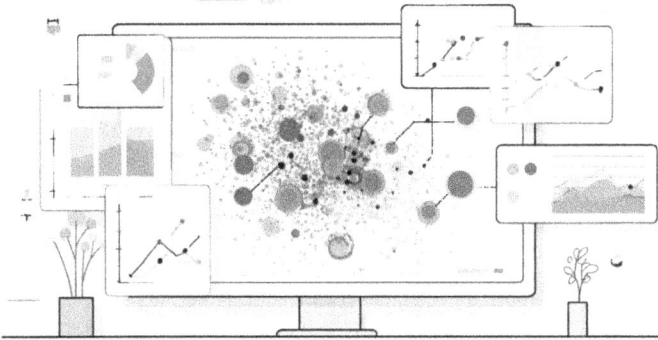

To build and maintain customer trust, communication is key. Being transparent about how AI works reassures customers. Giving clear, understandable explanations helps them feel confident in the system. And actively seeking and integrating customer feedback strengthens relationships and trust over time.

Audit trails play a big role in AI accountability. They document decisions and data use, making processes transparent. They also allow us to trace actions back to their source—helping detect errors and improve reliability. Plus, they support compliance by providing evidence that regulations and ethical standards are being met.

Strong governance standards are essential for responsible AI. Ethical principles ensure fairness and respect for rights. Risk management frameworks help identify and reduce potential harms. Data management policies protect the quality and security of data. And continuous evaluation keeps systems safe, fair, and compliant over time.

Monitoring and documentation are crucial for AI success. Regular monitoring ensures performance and ongoing compliance. Comprehensive documentation supports transparency, helps with audits, and drives continuous improvement of AI systems.

In conclusion, responsible AI use requires a balance of privacy, transparency, and governance. By understanding and applying GDPR, fostering explainability and trust, and maintaining strong governance and documentation, we can ensure that AI serves both businesses and customers ethically and effectively.

Chapter 16: AI-Augmented Vs Fully Autonomous Service Models

AI-augmented service models are designed to support human agents rather than replace them. They provide real-time insights, helping agents make better decisions in the moment. With decision support tools, AI suggests options or next steps, allowing human agents to respond more efficiently. Automation tools take care of routine tasks, freeing up the agents to focus on complex customer needs that require human empathy and judgment.

Fully autonomous service models take a different approach. These systems handle customer interactions entirely on their own, without human intervention. They manage everything from initial contact to resolution, using advanced AI capabilities like natural language understanding and machine learning to operate independently.

When comparing AI-augmented and fully autonomous models, each has its strengths and challenges. AI-augmented models excel at delivering personalized service but rely on human input, making them ideal for complex or sensitive interactions. Fully autonomous models offer unmatched scalability and can operate around the clock, though they face hurdles like complexity in deployment and customer trust. Use cases for AI-augmented models include bespoke customer support, while autonomous models fit high-volume, standardized operations.

Digital humans are virtual customer service representatives that simulate human-like interactions. They combine natural language processing with visual avatars, creating a more personal and engaging experience for customers. These digital humans can assist with routine queries, guide users through processes, and enhance service availability.

Emotion AI helps machines recognize and respond to human emotions. By analyzing voice tone, facial expressions, and language cues, AI systems can detect how a customer feels. This allows the system to tailor its responses with empathy, improving customer satisfaction. Emotion AI doesn't just react; it helps create a more human-like, supportive interaction that strengthens the customer relationship.

When emotion AI is used effectively, it can significantly boost customer satisfaction and engagement. Customers feel more understood and valued when their emotional state is acknowledged. This leads to better service experiences and deeper connections between customers and organizations.

AI agents can remember past interactions with customers, which allows them to provide more relevant and personalized service. They securely store customer data, ensuring that personal information is protected while enabling seamless support in future interactions.

By recalling customer preferences and previous issues, AI agents can tailor responses, offering personalized solutions quickly. This not only improves the efficiency of support but also makes the customer feel recognized and appreciated.

AI's ability to analyze past customer behavior also enables proactive support. Instead of just reacting, AI agents can anticipate needs, offer relevant suggestions, and even solve problems before the customer reaches out.

Personalization powered by AI leads to higher customer satisfaction. When customers feel understood, they are more likely to have positive service experiences. This tailored approach also speeds up resolution times, as AI can quickly access relevant information. Over time, these personalized interactions help build strong, lasting customer loyalty.

However, with personalization comes the responsibility of protecting customer data. Strict privacy measures must be in place to safeguard sensitive information. Transparent data policies help build trust by clearly explaining how customer data is handled. Ethical data use ensures that personalization enhances customer experience without compromising their privacy or rights.

Autonomous escalation processes involve AI identifying cases that require higher-level intervention. The AI can detect when an issue needs special attention and automatically route it to the right person or department, all without manual handling. This ensures that critical issues are addressed quickly and efficiently.

Several technologies support autonomous resolution. AI-powered chatbots can handle basic queries and even some complex issues independently. Machine learning models enable AI to make informed decisions based on data analysis. Robotic process automation takes care of repetitive tasks, improving service speed and accuracy.

Evaluating how well autonomous systems work is essential. We measure success by looking at resolution speed—how fast issues are solved—and the accuracy of those resolutions. Customer satisfaction is another key metric, helping us understand how these systems are performing from the user's perspective. Continuous learning allows AI systems to get better over time, refining their responses and improving service quality.

To conclude, both AI-augmented and fully autonomous service models are transforming customer experiences in unique ways. Whether through supportive tools, empathetic digital humans, personalized agents, or autonomous resolution processes, AI is reshaping how organizations interact with customers. The key to success lies in balancing innovation with ethical considerations, always keeping the customer at the center of service design.

Chapter 17: Service Architecture And Ecosystem Integration

AI integration with CRM systems delivers several key benefits: It allows for greater customer personalization, helping businesses tailor experiences and increase engagement. AI automates routine tasks, boosting efficiency and allowing teams to focus on strategic work. With advanced data analytics, AI helps extract actionable insights to improve decision-making. Finally, AI supports proactive customer service by anticipating needs and addressing issues before they arise.

Let's look at some examples of AI-powered automation in customer service: Chatbots can handle common inquiries and provide real-time support by simulating human conversation. Voice assistants offer hands-free service access, improving convenience and accessibility. Predictive analytics anticipate customer needs by spotting patterns in data, enabling more personalized service. Automated case routing ensures customer issues are directed quickly to the right department, reducing resolution times.

When integrating AI with CRM systems, there are a few key challenges and best practices to keep in mind: Poor data quality can severely impact AI effectiveness, so maintaining clean, accurate data is critical. System compatibility is another concern — AI tools need to integrate smoothly with existing CRM platforms. User adoption can also be a barrier, so providing training and ensuring ease of use is essential. Best practices include a phased rollout, collaboration across departments, and continuous monitoring to refine AI integration.

Omnichannel customer journeys often involve multiple interaction points across digital and physical channels. These paths can be complex, so it's important to understand how customers move between channels to deliver a seamless experience. Providing a cohesive customer experience means gaining insight into every interaction, no matter where it happens.

AI plays a crucial role in enabling seamless cross-channel experiences by: Analyzing data in real time to provide current insights across platforms. Offering dynamic personalization based on user behavior and preferences. Engaging customers in a context-aware manner, delivering relevant messages that resonate. Using predictive insights to forecast needs and behaviors, allowing businesses to proactively enhance experiences.

There are powerful case studies showing how AI drives omnichannel success: AI enables highly personalized interactions, increasing customer satisfaction and loyalty. Seamless omnichannel experiences foster deeper customer engagement. Ultimately, organizations using AI-driven omnichannel strategies see improved operational efficiency and increased revenue.

Building an AI-first customer service strategy starts with strong foundations: A robust data infrastructure is vital for supporting AI applications. AI technologies need to be scalable to handle business growth and data expansion. Skilled teams with AI expertise are critical for innovation and successful deployment. A customer-centric culture ensures that AI initiatives deliver real value and meet customer needs.

To align AI capabilities with business goals: Make sure AI tools and systems directly support strategic objectives. Use AI insights to boost customer retention by delivering more satisfying, personalized experiences. Leverage AI for cost reduction and service quality improvement through automation and better analytics.

Measuring the success of AI-first strategies involves several key steps: Define relevant KPIs to track performance effectively. Monitor AI system performance regularly to catch issues and optimize results. Integrate customer feedback to ensure solutions remain user-focused. Continuously refine AI models to enhance their accuracy and effectiveness over time.

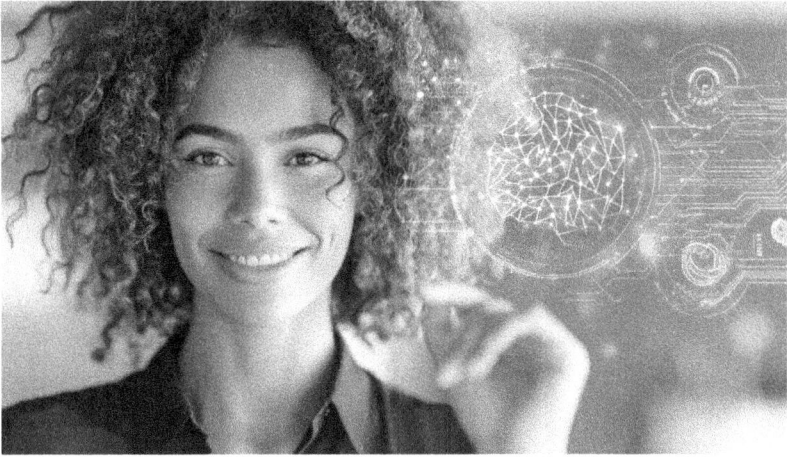

To wrap up, AI is reshaping customer service by enabling smarter interactions, seamless omnichannel experiences, and data-driven strategies. By integrating AI thoughtfully and continuously refining your approach, you can enhance customer satisfaction, improve operational efficiency, and drive meaningful business outcomes.

www.ingramcontent.com/pod-product-compliance
Lightning Source LLC
Chambersburg PA
CBHW060632210326

41520CB00010B/1571